A·C·T·I·O·N B·O·O·K

FLY THE
SPACE SHUTTLE

Carole Stott

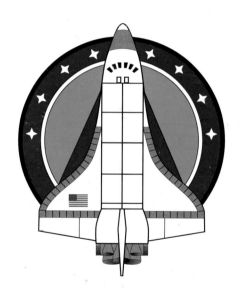

DORLING KINDERSLEY
LONDON • NEW YORK • MOSCOW • SYDNEY

A DORLING KINDERSLEY BOOK

Project Editor Claire Bampton
Editor Francesca Stich
Senior Editor Sue Leonard
Art Editors Jane Bull, Cathy Chesson, Vicky Wharton
Picture Researcher Monica Allende
Photographers Andy Crawford, Gary Ombler
Paper Engineer David Hawcock

Illustrator Hans Jenssen

Managing Editor Sarah Phillips
Senior Managing Art Editor Peter Bailey
DTP Designers Karen Nettelfield, Andrew O'Brien
Design Assistant Miranda May
Production Joanne Blackmore, Lauren Britton

First published in Great Britain in 1997
by Dorling Kindersley Limited
9 Henrietta Street, London WC2E 8PS

Copyright © 1997 Dorling Kindersley Limited, London

Visit us on the World Wide Web at http://www.dk.com

A CIP catalogue record for this book is available
from the British Library.

ISBN: 0-7513-5625-5

Colour reproduction by Flying Colours, Italy
Manufactured in China for Imago

Identity badge

There are hundreds of people
working at the Space Centre.
Everyone wears a badge so you
know instantly who they are and
what their role is. Here is your
badge. Please add your details
and photograph to it now, so
that your training can begin.

In the Space
Centre you should
display your badge
at all times.

Space Centre

Glue a
photograph of
yourself in here.

Name ....

Age ....

Height ....

Signature ....

Training Tasks

Your training is divided into
four modules, each made up of
a number of training sessions.
Each module is taught by an
instructor who will set you a
training task at the end of each
session. Work through the
training tasks carefully on a
separate piece of paper. You
can then check whether your
answers are correct on page 30.

The Space Shuttle will
transport you into space.

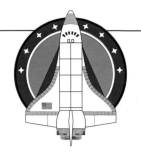

Space Centre

Welcome to the Space Centre and congratulations on being chosen to join the latest class of trainee astronauts. You are one of the lucky few. Thousands of people apply to be Space Shuttle astronauts each year, but only a handful are chosen. You were selected because you are fit, educated, and have the personal qualities necessary to be a good crew member. At the Space Centre we are going to build on these qualities and develop the knowledge and skills you will need in space. The four key people who will guide and assist you through your four training modules will introduce themselves to you. Pay attention to what they have to say – the success of your mission may depend on it.

Good luck and, when the time comes, enjoy your space trip!

MODULE CONTENTS

Steve Harrison –
Space Coordinator

David Bridges –
Astronaut Trainer

Ellen Collins –
Scientific Advisor

Owen Mitchell –
Space Shuttle Pilot

MODULE A
In this module you will learn what space is like, how far you will be travelling into space, and find out about the vehicle in which you will travel.

MODULE B
In this module you will be issued with your kit, complete your practical training in the simulators, and practise the emergency procedures.

MODULE C
In this module you will be briefed about the launch and find out what it is really like to live and work in space.

MODULE D
In this module you will learn about space stations, how to dock with them in the Space Shuttle, and how to make a successful return home.

Destination Space

"Hello, I'm Steve Harrison, Space Coordinator, and I'm here to start you on your journey to space. We'll be working together for 'Module A' of your astronaut training."

■ To gain astronaut status you must travel at least 80 km (50 miles) above Earth. That is not far – it takes about eight minutes in the Space Shuttle. However, as you leave Earth and move through its atmosphere, the environment changes and becomes deadly to humans. In this training session we will compare Earth with space, discuss how the Space Shuttle is used, and look at its work with other spacecraft.

Life on Earth

The conditions on Earth suit you perfectly. The air contains oxygen for breathing, the pressure of the air allows you to breathe easily, and your surroundings are neither too hot nor too cold. At Earth's surface the pull of gravity keeps your feet firmly on the ground, but you can still move about with ease.

AIR	78% nitrogen, 21% oxygen, 1% other gases
TEMPERATURE RANGE	-70 to 55° C (-94 to 131° F)

Earth's surface

Stratosphere – between 11 and 50 km (7 and 31 miles) high

Troposphere – up to 11 km (7 miles) high

Mesosphere – between 50 and 80 km (31 and 50 miles) high

Thermosphere – between 80 and 480 km (50 and 298 miles) high

Meteor shower

Helium-filled balloon

Glider

Jumbo jet

Mount Everest

SPACE HIGHWAY

Every year about a hundred craft are launched into Earth's atmosphere and beyond. Some transport people, some collect and transmit information, some carry out experiments, and some observe the solar system. How far they go above Earth depends on the type of job they are doing.

Life in space

You cannot survive in space without support and protection. In space there is no oxygen to breathe, the temperature is either hot enough to turn you into toast or cold enough to freeze you solid, and the air pressure is so low that your body liquids evaporate. The pull of gravity is weaker than it is on Earth.

AIR	No air
TEMPERATURE RANGE	-101 to 121° C (-150 to 250° F)

The Russian space station, *Mir*, has been manned by astronauts since February 1987 and is about 400 km (249 miles) above Earth.

Satellites are about 250 to 1,000 km (155 to 621 miles) high as they travel around Earth.

Gravity to weightlessness

The further you travel above Earth, the further you move away from Earth's gravity. If a spacecraft does not travel far enough to get away completely, it will be held in an orbit around Earth. As the Shuttle orbits, the surface curves away beneath it, so it is constantly falling around the Earth. This makes it and everything in it feel weightless.

Earth's atmosphere

Earth's atmosphere is made up of five main layers. The troposphere is the nearest layer. Most of the weather occurs here. Next comes the stratosphere, which is a calm region. The mesosphere can be very cold, -100° C (-148° F). In the next layer, the thermosphere, the air is thin and very hot.

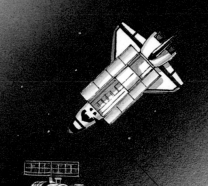

Into space

There is no neat dividing line between the exosphere and space – one fades into the other. This is where you travel in the Space Shuttle.

Planning an orbit

The route you take around Earth for your mission is worked out well before your date of departure. But whatever distance you are from Earth, each day you travel from brilliant sunny skies to total darkness.

TASK 1

Based on the information given about the Space Shuttle and its orbit in the diagram, work out how long it takes to complete one orbit of Earth.

TIP
Remember to convert your calculations into hours and minutes.

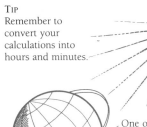

The Space Shuttle is travelling at an average speed of 26,548.5 kph (16,486.5 mph).

One orbit of Earth is about 39,823 km (24,750 miles).

YOU ARE HERE
The Space Shuttle flies between 200 and 1,000 km (124 and 621 miles) above Earth's surface.

Exosphere – between 480 and 1,000 km (298 and 621 miles) high

The *Hubble Space Telescope* travels around Earth at an average height of 500 km (311 miles).

TASK 2

Using your answer to Task 1, work out how many times you see the Sun rise in 24 hours (one Earth day).

OBSERVATION
Before you fly into space, it may be possible for you to spot the Space Shuttle in orbit. Observe the early morning or early evening sky. The Space Shuttle looks like a moving star.

SPACE SHUTTLE IN ACTION

The Space Shuttle is capable of carrying out a number of tasks, from scientific experiments to launching other craft, and your mission could involve any one of them.

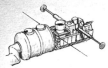

Sometimes the Space Shuttle carries a mobile laboratory called Spacelab. It consists of a laboratory and pallets full of experiments.

Often the Space Shuttle is used to release satellites into space. It can also be used to capture and repair them.

The *Hubble Space Telescope* was launched by the Space Shuttle. Later crews also repaired the telescope while in space.

Space probes investigate our solar system. The Space Shuttle is used as a launch pad to send them into space.

The Space Shuttle can dock with space stations, like *Mir*, to deliver and collect astronauts and supplies. It can also be used to build space stations.

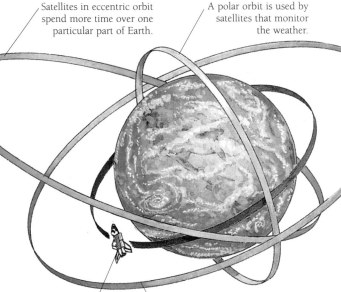

Satellites in eccentric orbit spend more time over one particular part of Earth.

A polar orbit is used by satellites that monitor the weather.

The Space Shuttle travels in low Earth orbit.

A satellite in geostationary orbit travels at the speed at which Earth is spinning. This means it stays over the same part of Earth during its complete orbit.

Orbiting Earth

Spacecraft orbit, or circle around, Earth. The orbit they take determines the distance and the speed at which they travel. There are four main orbital paths: polar orbit, low Earth orbit, eccentric orbit, and geostationary orbit. The Space Shuttle follows a path in low Earth orbit.

Vehicle for Space

■ The Space Shuttle is the only successful reusable space vehicle in the world. Like other spacecraft, the Space Shuttle is boosted into space by its rockets. But unlike any other craft, it glides back to Earth to be used again by a new crew of astronauts. After the completion of a mission, the Space Shuttle can be reassembled for another launch in just six and a half days.

THE TOTAL SYSTEM

The Space Shuttle is made up of three parts: a reusable plane, called an orbiter, two solid fuel rockets, which are recovered and reused on later missions, and an expendable (throwaway) fuel tank.

The assembled vehicle is 56 m (184 ft) high and 24 m (78 ft) wide.

The fuel tank supplies the orbiter's main engines during blastoff and ascent.

The two solid fuel booster rockets propel the Space Shuttle away from Earth.

The orbiter is about the size of a small aeroplane.

The Space Shuttle is assembled in the vehicle assembly building – one of the largest buildings in the world.

FLIGHT PROFILE

The complete Space Shuttle system is together for only a few minutes at the start of your journey. Once the fuel tank is empty and the rockets have burned out, they both fall off leaving you and the rest of the crew in the orbiter to complete your mission.

4. The two solid fuel rockets are discarded.

5. The fuel tank is released, leaving only the orbiter.

6. The orbiter's on-board engines are used to reach low Earth orbit.

7. Astronauts work inside and outside the orbiter as it travels around Earth.

8. The orbiter positions itself ready to return to Earth.

The fuel tank falls back into Earth's atmosphere and breaks up over the ocean.

As the rockets fall back to Earth their parachutes open.

3. The Space Shuttle blasts off and moves upwards, away from the launch pad.

Waiting ships recover the two rockets from the ocean.

1. The Space Shuttle is made ready in the vehicle assembly building.

2. The Space Shuttle travels on the crawler transporter to the launch pad.

Launch to landing

Your trip in space will last four to seven days. As you travel, some ground crew will retrieve the reusable parts of the Space Shuttle, while others will monitor your progress in space. Everyone will be on stand-by for your return.

9. The orbiter moves downwards through Earth's atmosphere.

10. The orbiter gets ready for a high-speed glide onto the runway.

11. The orbiter's landing gear is lowered.

12. The orbiter glides in to land on the extra-long runway.

Full of fuel

The external fuel tank is the largest and heaviest part of the Space Shuttle. During blastoff it burns 242,240 litres (53,285 gallons) of fuel a minute.

The fuel tank is 47 m (154 ft) long and 8.4 m (27.6 ft) in diameter.

Liquid oxygen tank

Liquid oxygen fuel pipe leading to orbiter

Liquid hydrogen tank

Liquid hydrogen fuel pipe leading to orbiter

The rocket ignites here.

On ignition the fuel burns from top to bottom in all four segments of each rocket.

Each rocket is 45.5 m (149.2 ft) long and 3.7 m (12.2 ft) in diameter.

The four central segments contain the solid fuel.

Rocket power

The two solid fuel rocket boosters have to be powerful enough to lift the system off the ground. When the fuel is lit, hot gases are forced downwards through a nozzle. This propels the Space Shuttle in the opposite direction.

Hot gases escape from the rocket, forcing the Space Shuttle upwards.

Escaping gravity

The Space Shuttle's speed increases quickly as it moves away from Earth. It needs to go fast to get away from gravity's pull and achieve its low Earth orbit. The speed, or orbital velocity, needed for this is 7.9 kps (4.9 mps). The Space Shuttle's orbit is circular so its speed is constant. The speed of a spacecraft on an eccentric orbit varies. It moves fastest when closest to Earth.

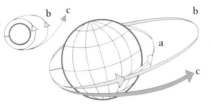

a – low Earth orbit

b – eccentric orbit

c – start of distant orbit

TASK 1

Using the diagrams work out which is the Space Shuttle's orbit, **a** or **b**.

TASK 2

Would the orbital velocity of a spacecraft on path **b** be fastest or slowest when furthest away from Earth?

TASK 3

Is path **c** the start of a circular or an eccentric orbit?

The parachutes are stored in the nose.

Small rockets are used to separate the rocket from the Space Shuttle.

The crawler moves at about 1.6 kph (1 mph).

The crawler transporter is 6.1 m (20 ft) high, 40 m (131 ft) long, and 34.7 m (114 ft) wide.

The orbiter, rockets, and fuel tank are fixed together for blastoff.

Pipes deliver fuel from the fuel tank to the orbiter.

During liftoff and ascent the orbiter is powered by fuel from the external fuel tank.

Small rockets are fired to separate the rocket from the Space Shuttle.

Ready for launch

The Space Shuttle is put together on the crawler transporter, the largest land vehicle in the world. Slowly the crawler transporter moves towards the launch pad.

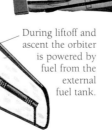

Piggyback transport

The orbiter is not always launched from its last landing site. A specially adapted Boeing 747 aircraft can carry the orbiter on its back between different sites.

The Orbiter Space Plane

■ The Space Shuttle's orbiter is one of the most amazing vehicles ever built. About the size of a small aeroplane, it contains everything you and the rest of the crew will need to survive for the duration of your mission, and is fully equipped to carry out the necessary space work. In this training session we will look at the orbiter, how it is powered, and how to operate it.

ORBITER FLEET

You will travel in one of four orbiters: *Columbia*, *Discovery*, *Atlantis*, or *Endeavour*. They are all identical in size and structure.

LENGTH	37 m (122 ft)
HEIGHT	17.4 m (57 ft)
WINGSPAN	24 m (78 ft)

Top view

Side view

WEIGHT	84,778 kg (187,000 lb)
CARGO CAPACITY	22,680 kg (50,000 lb)

Bottom view

Payload bay

The large cargo area in the orbiter is called the payload bay. It is 18.3 m (60 ft) long by 4.6 m (15 ft) wide. Satellites, space probes, or Spacelab are carried up to space inside the payload bay.

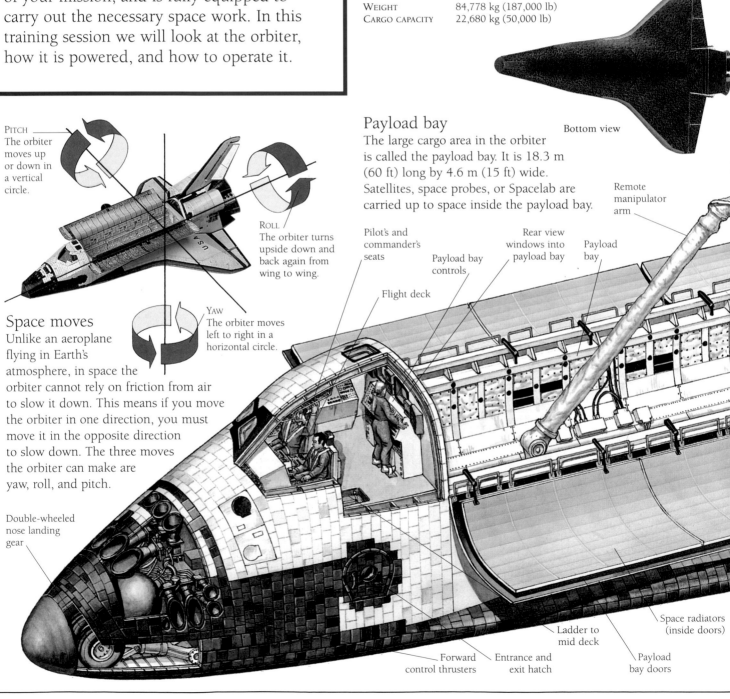

PITCH
The orbiter moves up or down in a vertical circle.

ROLL
The orbiter turns upside down and back again from wing to wing.

YAW
The orbiter moves left to right in a horizontal circle.

Space moves

Unlike an aeroplane flying in Earth's atmosphere, in space the orbiter cannot rely on friction from air to slow it down. This means if you move the orbiter in one direction, you must move it in the opposite direction to slow down. The three moves the orbiter can make are yaw, roll, and pitch.

Double-wheeled nose landing gear

Pilot's and commander's seats

Payload bay controls

Rear view windows into payload bay

Payload bay

Remote manipulator arm

Flight deck

Forward control thrusters

Entrance and exit hatch

Ladder to mid deck

Space radiators (inside doors)

Payload bay doors

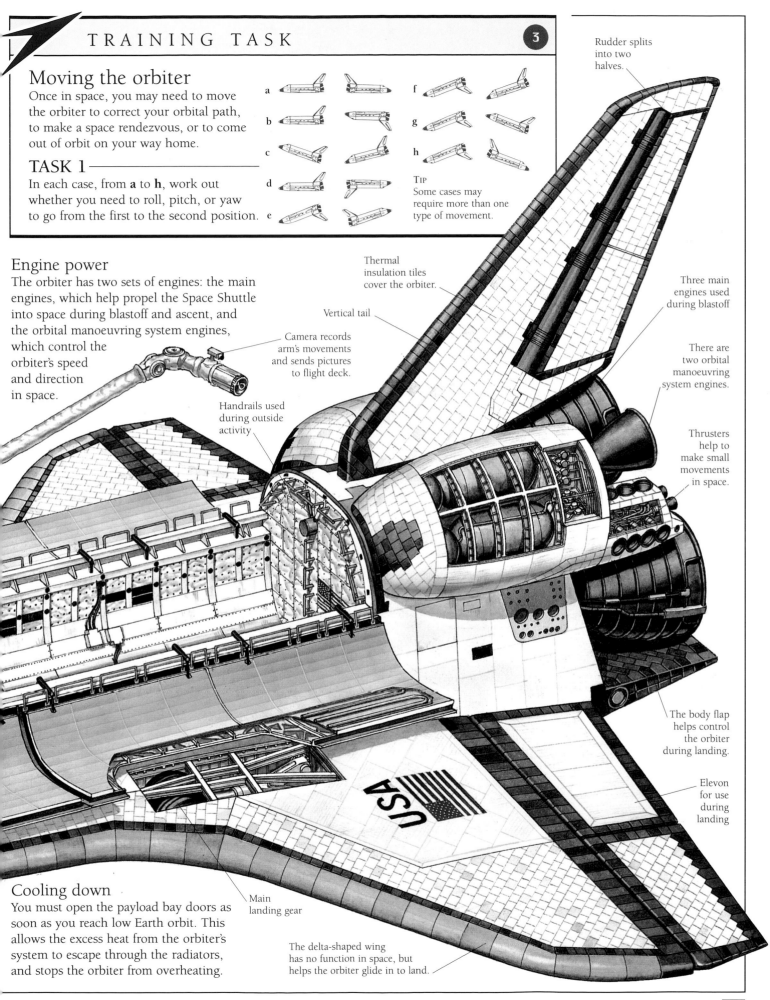

Moving the orbiter

Once in space, you may need to move the orbiter to correct your orbital path, to make a space rendezvous, or to come out of orbit on your way home.

TASK 1

In each case, from **a** to **h**, work out whether you need to roll, pitch, or yaw to go from the first to the second position.

a

b

c

d

e

f

g

h

TIP
Some cases may require more than one type of movement.

Engine power

The orbiter has two sets of engines: the main engines, which help propel the Space Shuttle into space during blastoff and ascent, and the orbital manoeuvring system engines, which control the orbiter's speed and direction in space.

Camera records arm's movements and sends pictures to flight deck.

Handrails used during outside activity

Thermal insulation tiles cover the orbiter.

Vertical tail

Rudder splits into two halves.

Three main engines used during blastoff

There are two orbital manoeuvring system engines.

Thrusters help to make small movements in space.

The body flap helps control the orbiter during landing.

Elevon for use during landing

Main landing gear

The delta-shaped wing has no function in space, but helps the orbiter glide in to land.

Cooling down

You must open the payload bay doors as soon as you reach low Earth orbit. This allows the excess heat from the orbiter's system to escape through the radiators, and stops the orbiter from overheating.

Selecting your Kit

"My name is David Bridges. We'll be working together for 'Module B'. I'll help you prepare for everyday life in space and show you how to overcome the problems you may encounter."

■ In this training session we will look at the personal items that you will be given for use during your space flight. These include clothing, crew equipment, and sighting aids, and are either packed away in the orbiter, or given to you before the takeoff. Pay attention, because each item has been carefully designed and what you wear and use depends on the type of job you are doing.

Packed up
All your personal items will be packed in storage trays before takeoff and secured in lockers in the mid deck ready for use.

Also packed away will be special-issue underwear – the longer your flight, the more changes you will get!

CREW CLOTHES
There are three types of space outfit: a partial pressure suit for launch and return, in-flight garments for everyday orbiter life, and an extravehicular mobility unit (EMU) spacesuit to protect you if you go outside the orbiter.

Partial pressure suit
This one-piece suit has an inflatable lower half that stops blood from pooling in the lower half of your body. This reduces the danger of you fainting as the spacecraft returns from weightlessness to Earth's gravity. The suit also has a parachute harness and pack, which are needed if an emergency arises and you have to leave the orbiter.

You will be helped into your suit by a technician before you enter the orbiter.

The badge on your left breast displays your name.

The suit's bright colour makes you easy to spot for rescue.

In-flight garments
After the ascent, you can change into your in-flight garments. These include trousers, lined zipper jackets, knit shirts, and sleep shorts. They have to be made of materials that are both comfortable and flame retardant.

Flapped pockets cover the outside of the trousers and jackets and are used for storing items such as pens, a pocket knife, a torch, food sticks, sunglasses, and scissors.

All the clothes come in three different sizes: small, medium, and large.

Footwear
There is no need for shoes inside the orbiter. These slipper socks with soft, thickened soles will keep your feet warm.

Getting dressed

The EMU spacesuit is kept in the airlock and this is where you will put it on. Getting into the spacesuit can be tricky, and will probably take about 15 minutes.

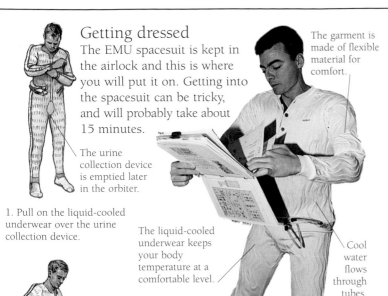

The urine collection device is emptied later in the orbiter.

The garment is made of flexible material for comfort.

The liquid-cooled underwear keeps your body temperature at a comfortable level.

Cool water flows through tubes.

1. Pull on the liquid-cooled underwear over the urine collection device.

2. The spacesuit is in two pieces. First pull on the trouser part with built-in shoes.

EMU SPACESUIT

The EMU spacesuit, for working outside, provides oxygen for you to breathe and keeps the air pressure around your body at the right level. The suit also protects you from the extreme heat and cold of space, and from fast moving space dust.

The helmet's visor provides protection against ultraviolet and infrared radiation from the Sun.

The primary life-support system, or PLSS, backpack contains oxygen for breathing.

3. The top half of the suit hangs on the airlock wall. Put it on by moving up into it.

The gloves are the only item of clothing made to your specific requirements.

The "Snoopy Cap" contains headphones and a microphone for two-way communication.

Bearings in the shoulder, arm, wrist, and waist joints allow you to move with ease.

4. Fit the two halves tightly together. Next, put on the communications cap, gloves, and finally the helmet.

Crew equipment

There are lots of useful items packed into storage lockers on the orbiter for everyone to use. There is also a selection of cameras for taking photographs and video films. Use them to record your trip.

On-board crew equipment:
- Torches
- Scissors
- Pocket food
- Plastic mirrors
- Pens and pencils
- Swiss army knives
- Sunglasses and pouches
- Emergency medical kit
- Personal hygiene kits
- Watches and watchbands

On-board sighting aids:
- Cameras
- Video cameras
- Binoculars
- Star and Earth maps

You can take a personal stereo and your favourite music CD if you wish.

Personal items

There may be something you wish to take on your trip. Everything must be approved, but if it is small, light, and not a fire hazard you will probably be allowed to take it.

TRAINING TASK ④

Taking photos in space

The success of your photographs depends on the amount of light reaching the film. You will not need so much light for close-up photographs as you will for long-distance photographs. You can control the amount of light entering the camera by altering the aperture (front opening), and the length of time it is open. The aperture is indicated by a series of f numbers – the smaller the number, the larger the aperture, and the shutter time can be set to be open for fractions of a second.

TASK 1

Which of the settings, **a** or **b**, would you use for: 1) a close up of an astronaut eating inside the orbiter, and 2) a long–distance view of your home town on Earth?

a) 5.6 and 1/60
b) 5.6 and 1/500

TIP
The f number range is: 1.4, 2, 2.8, 4, 5.6, 8, 11, 16, 22. The exposure times are: 1/60, 1/125, 1/250, 1/500, 1/1000 of a second.

Astronaut Training

■ During this training session you will learn about the mission simulators. These help you to become familiar with the practical aspects of your journey without having to travel to space. In the months before takeoff you will go through the details of your mission, from launch to landing, using computer software. You will practise these mission procedures over and over again until they become second nature.

MISSION SIMULATORS

Simulators are large pieces of training equipment that are built to re-create the conditions you will find in space. They include exact copies of sections of the orbiter.

Orbiter simulators
The two orbiter simulators each consist of the nose end of an orbiter and are positioned as if ready for launch. One of them is fixed, but the other moves to simulate takeoff.

Inside the orbiter simulator you lie in a flight chair ready for "takeoff"

Images of Earth from space or the landing runway are projected onto the windows, so that you see exactly what astonauts see during a real space trip.

Displays and controls are the same as in the real orbiter.

Remote manipulator arm

You can operate the manipulator arm to move cargo in and out of the payload bay.

Flight-deck controls
In the flight deck mock-up you can practise using the controls for all parts of the flight – prelaunch, ascent, orbit operations, reentry, and landing.

Working practice
You can practise payload bay procedures using these mock-ups of the payload bay and the flight deck. They help you become familiar with the remote manipulator system.

Life-size orbiter payload bay

The commander (left) and pilot (right) use the flight deck mock-up to practise landing on the runway.

A large drum-shaped helium balloon simulates a satellite cargo.

You can look through the flight deck windows into the payload bay and watch the manipulator arm move.

Going up

Coming down

You will experience weightlessness like this about 40 times in one day.

Training for weightlessness
For all but the very start and end of your trip you will feel weightless. To get used to this you can travel in the KC-135 aircraft. The seatless airliner flies to a height of 10.5 km (6.5 miles), then quickly descends to 7.3 km (4.5 miles). During this descent you will be weightless for about 20 seconds.

Repeat practice
If your particular mission involves any extravehicular activity you will practise performing the tasks, in an EMU suit, until you feel as though you could do them in your sleep.

Weightless-environment training facility
When you are underwater the effect of gravity on your body is lessened, as it is in space, so this deep pool of water is where you practise working in a weightless environment. The submerged payload bay allows you to practise your specific EVA jobs. Wearing an EMU suit you can attach yourself to the remote manipulator arm, then carry out repairs using the tools you will work with in space.

Gravity and weight
You have weight because Earth's gravity pulls on you. An increase in gravity's pull makes you heavier, a decrease makes you lighter. As you move away from Earth towards space, gravity's pull is lessened and you become lighter.

TASK 1
Imagine you could stand at different heights above Earth. If you weighed 100 kg (220.5 lb) on Earth (at **a**), work out your weight at positions **b**, **c**, and **d**.

TIP
To work out your weight, divide your weight on Earth by the square of the radial distance from Earth's centre. (The radial distance is the distance from the centre of a sphere to any point on its surface.)

d   Radial distance from Earth's centre is 2.5.

c   Radial distance from Earth's centre is 2.

b   Radial distance from Earth's centre is 1.5.

a   Radial distance from Earth's centre is 1.

Earth

Surviving in Space

■ In this training session we will look at some of the problems you may encounter in space. They could be to do with you and how your body reacts to space life, or they might be because of a malfunction with the Space Shuttle or one of its systems. Whatever the nature of the problem, a solution procedure has been formed for it and ground control and the rest of the crew will be on hand to help.

EMERGENCIES NEAR EARTH

It is possible that a problem may occur in Earth's environment, either as you wait strapped in your seat on the launch pad, or during takeoff and landing.

Rapid exit

If you are travelling through Earth's atmosphere and cannot land, use the side hatch in the mid deck to exit the craft. First adjust the deck pressure so that it is the same as the pressure outside. Then put the escape pole through the side hatch, slide down the pole, and let go at the end.

The parachute in your partial pressure suit opens automatically.

The whole crew can bail out in less than 90 seconds in this way.

SPACE SICKNESS

Most astronauts feel sick during the first few days in space. In a weightless environment your internal organs and body fluids shift position as there is no gravity pulling them downwards. This can cause nausea, headaches, sweating, and vomiting.

At the start of the trip your nose is blocked because your sinuses are congested with extra fluid.

Your face looks puffy as fluids move upwards.

In space your heart does not need to pump as hard as it does on Earth and can get weak.

Your bones become weaker as minerals such as calcium are lost.

The weightlessness in space makes you "stand" on your toes with your knees bent, known as the "bird-legs" effect.

Your kidneys get rid of about 2 litres (3.5 pints) of fluid in the first few days.

In space you are 1-2 cm (0.3-0.75 in) taller than you are on Earth because the vertebrae in your spine float apart.

Steel baskets

In the 30 seconds before countdown ends and access to the launch tower is removed, you can escape from the orbiter in one of the seven steel baskets attached to the tower.

Armoured vehicle

If you need to abandon the orbiter before takeoff, an armoured vehicle will transport you away from the launch pad. It can move across rough terrain to take the shortest route to safety.

The armoured vehicle protects you from an explosion.

SPACE EMERGENCIES

Lights on the control panels, and alarms that sound throughout the orbiter and in your headset, will alert you if something unusual happens during your space flight.

Payload doors

One of the problems you may have is closing the payload doors. If the automatic closing system fails, you will have to close the doors manually. Put on your EMU suit and go outside to assess the situation. Discuss the problem with the crew and ground control before carrying out the correction procedure.

The orbiter cannot return to Earth until the payload doors are closed.

Abandon ship

If something goes seriously wrong, you may have to abandon the orbiter and transfer to another craft. If this happens, put on either an EMU suit or a partial pressure suit. The EMU suit has an oxygen supply and provides protection until you are safe in the second craft.

Monitoring your body

You must remain fit in space and be aware of your state of health. However, to know whether you are healthy or not, you need to know what is normal. Answer the questions below to assess what you know about the human body.

TASK 1

1. What is the normal temperature of a healthy human being?
a) 35° C (95° F) b) 37° C (98.6° F) c) 39° C (102.2° F)

2. Using what you have learned so far about space, what do you think might happen to your blood if you pricked your finger in the orbiter and it started to bleed?

3. Dust does not settle in space, but floats around inside the orbiter, gets in your nose, and makes you sneeze. How many times a day on average do you think you might sneeze in space? a) 5 b) 50 c) 100

Escape ball

When abandoning the orbiter in space, astronauts wearing partial pressure suits should get into a personal rescue enclosure. These 86-cm (34-inch) inflatable balls are pressurized and are just big enough for one person to sit inside cross-legged. The EMU-suited crew can then transfer the personal rescue enclosures to the second craft.

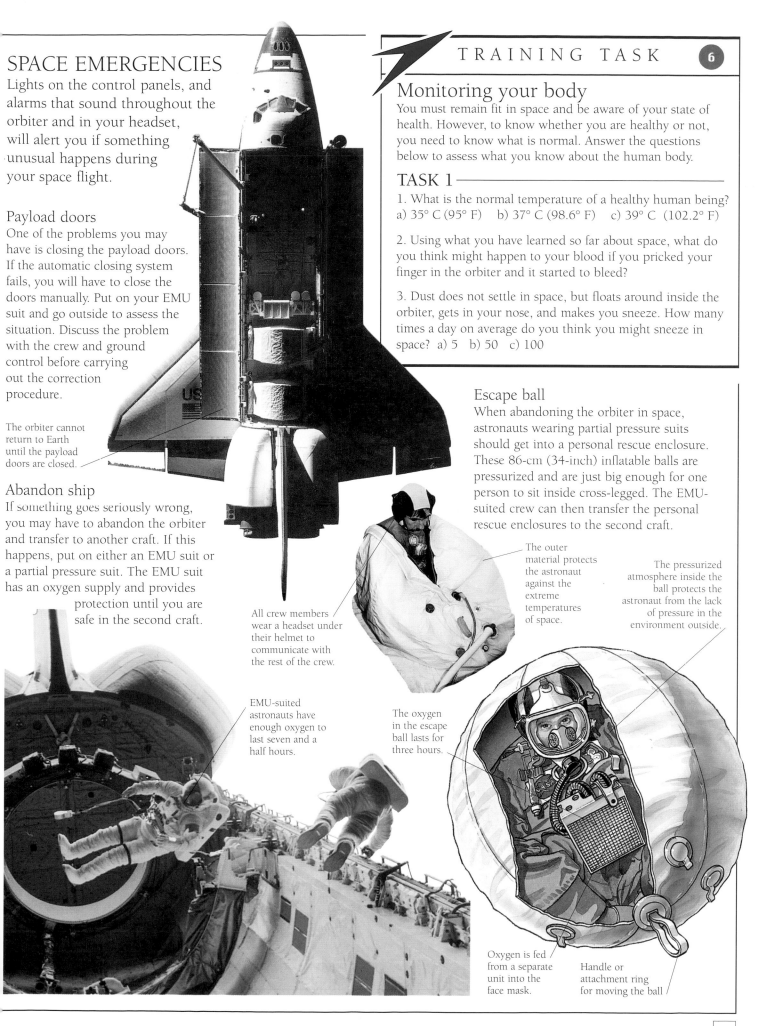

All crew members wear a headset under their helmet to communicate with the rest of the crew.

EMU-suited astronauts have enough oxygen to last seven and a half hours.

The outer material protects the astronaut against the extreme temperatures of space.

The pressurized atmosphere inside the ball protects the astronaut from the lack of pressure in the environment outside.

The oxygen in the escape ball lasts for three hours.

Oxygen is fed from a separate unit into the face mask.

Handle or attachment ring for moving the ball

Countdown and Blastoff

"Hello, I'm Ellen Collins, your Scientific Advisor and instructor for 'Module C'. I've flown on two Space Shuttle missions and can tell you what it is like to live, work, and travel in space."

■ In this training session, we will take a look at your launch into space. It is a really short part of your whole trip, but in the few minutes it takes, you will have lots of new experiences. Although you will have practised the launch hundreds of times in the orbiter simulator, this is never as exhilarating as the rush of noise, the vibrations, and the surge of motion that you encounter on your first real trip into space.

COUNTDOWN TO LAUNCH

TIME PRIOR TO TAKEOFF (T)	PROCEDURE	ACTION
hrs mins secs **5.00.00**	Begin launch day	Wake, breakfast, and dress. Meanwhile, fuel is pumped into the external fuel tank.
1.50.00	Enter the orbiter	Enter the orbiter and climb into your seats. Secure your safety straps.
1.30.00	Check communications with launch control	Launch control contacts you on air-to-ground intercom and air-to-air channels.
1.25.00	Check communications with mission control	Mission control contacts you on air-to-ground channels.
1.10.00	Close the entrance and exit hatch	Ground crew closes and secures the side hatch.
0.20.00	Load the Space Shuttle computer	Load the final mission data into the computer.
0.09.00	Go for automatic countdown	Reposition timer switch to "start". Automatic countdown begins.
0.00.30	Cut electrical links with Earth	Switch management of launch from launch control to on-board computers.

UP AND AWAY

Your launch into space is automatically controlled by on-board computers, but this does not mean that you have nothing to do. Your main job is to check everything runs smoothly and to solve any problems that may arise.

You hear the orbital manoeuvring system engines ignite and the orbiter moves into low Earth orbit.

You feel weightless and you can look out of the window and see Earth.

T +50 minutes
You are in low Earth orbit, about 400 km (250 miles) above Earth.

T +10 minutes 39 seconds
Your speed is 28,157.5 kph (17,500 mph) and your height is 273.5 km (170 miles).

Mission control

When you reach orbit, launch control passes you over to mission control, a team of people on Earth who monitor your every movement. You will be in constant contact with mission control for the rest of your flight.

Although you cannot see it, the empty fuel tank is ejected from the orbiter.

T +8 minutes 54 seconds
Your speed is 27,353 kph (17,000 mph) and your height is 117.5 km (73 miles).

View from space

There are two sure signs you have reached space. Firstly, if you loosen your straps you will start to float across the orbiter. If you can control yourself, head for a window. There you will find the second sign. The view is breathtaking and one you will never forget.

T +2 minutes 7 seconds
Your speed is 4,972 kph (3,098 mph) and your height is 51 km (32 miles).

Within thirty seconds you move out of Earth's blue sky into blackness.

You see a bright flash of light as the empty rockets are fired off.

The orbiter vibrates wildly as you lift off the launch pad.

T +11 seconds
Your speed is 123 kph (76 mph).

Takeoff!

As the Space Shuttle accelerates you are pushed back into your seat. The force is three times the strength of gravity, which means you suddenly feel three times as heavy as you do on Earth.

TRAINING TASK

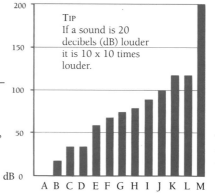

Space noise

Your takeoff will probably be the noisiest time of your life. A Space Shuttle launch is one hundred million times louder than a normal conversation. The loudness of sounds is measured in decibels. Each ten decibel increase means the sound is ten times louder.

TASK 1

Use the graph to work out how much louder a Space Shuttle launch is than:
a) normal traffic and b) classroom noise.

TASK 2

Space is silent because there is no air in which sound can travel. But inside the orbiter, air is pumped around the cabins, and life is pretty loud. Where do you think the noise inside the orbiter should go on the graph, position D, G, or I?

TIP
If a sound is 20 decibels (dB) louder it is 10 x 10 times louder.

KEY
A – No sound
B – Falling leaves
C – Normal conversation
D – ?
E – Busy office
F – Classroom noise
G – ?
H – Normal traffic
I – ?
J – Rock group
K – Space Shuttle launch
L – Noise painful to ears
M– Large rocket launch

Everyday Life in Space

■ Almost everything you do during a normal day on Earth, such as eating, drinking, sleeping, going to the toilet, and washing, you can do in space. You usually do these things automatically, but in the weightlessness of space you need to really think about your actions and the consequences of them. In this training session we will look at how best to perform everyday activities in space.

DOUBLE DECKER

Most of your time in space is spent inside the orbiter in the flight and mid decks. Together the decks contain everything you need to live and work during your mission. The decks actually feel larger than they look. In space you use the whole room, from floor to ceiling, not just the floor area as you do in rooms on Earth.

Your view of Earth from a flight deck window

Ship shape

In the weightless environment of space everything floats around with the slightest push. Items that are set loose become a potential hazard. This is why it is more important now than at any other time in your life to tidy up as you go along!

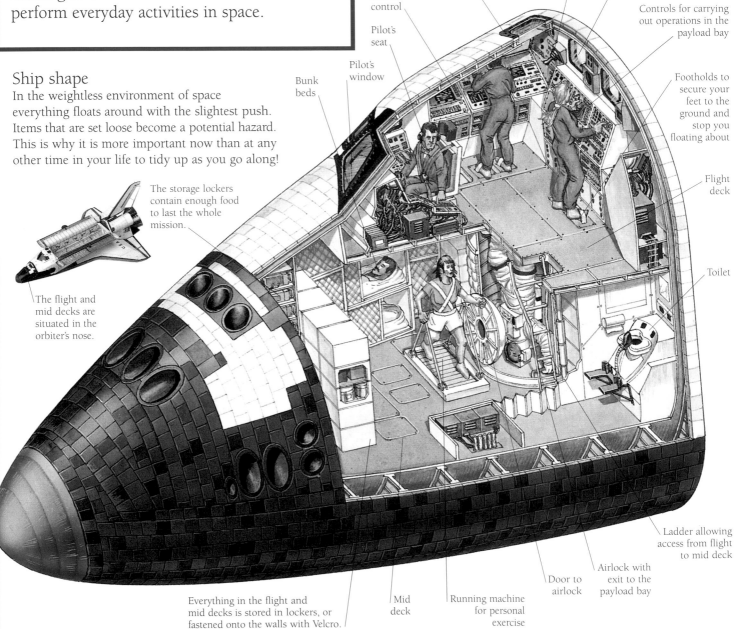

The storage lockers contain enough food to last the whole mission.

The flight and mid decks are situated in the orbiter's nose.

Controls used to contact mission control

Flight deck window

Joystick for controlling the remote manipulator arm

Window into payload bay

Controls for carrying out operations in the payload bay

Pilot's seat

Pilot's window

Bunk beds

Footholds to secure your feet to the ground and stop you floating about

Flight deck

Toilet

Ladder allowing access from flight to mid deck

Airlock with exit to the payload bay

Door to airlock

Running machine for personal exercise

Mid deck

Everything in the flight and mid decks is stored in lockers, or fastened onto the walls with Velcro.

A DAY IN THE ORBITER

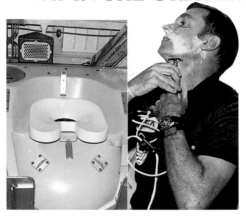

Secure the food and cutlery tray to your leg.

Dried fruit and nuts are ready to eat.

Foods may need to be warmed, to have water added, or may be ready for eating immediately.

Foods are in individual packages.

Drinks are usually taken from resealable containers, but you may have to test a can adapted for use in space.

07:00 hrs Brushing up
Using the toilet is easy – just secure yourself before you start! Use wet wipes to wash your face and hands. If a crew member needs to shave, he must make sure he vacuums up the cuttings before they start floating about.

07:30 hrs Flying food
You can use fingers, knives, forks, spoons, and chopsticks to eat. The main difficulty is trying to get the food from the tray into your mouth. Eating from one package at a time will help you avoid losing control of your food.

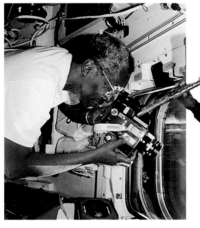

Small, light personal items, such as a flute, can be taken on your flight.

Your bunks have sheets with restraints to stop you floating off.

Footholds stop you drifting about.

10:00 hrs Running time
Weightlessness makes it easy to move around, so your heart and muscles do not work very hard. Two hours a day on the running machine keeps you fit.

19:00 hrs Leisure time
At the end of the day you will have some free time and it is up to you how you use it. You may want to take photos from the window, or you can just sit and stare if you wish.

23:00 hrs Bedtime
As you settle down in one of the enclosed bunk beds, another crew member gets up and starts the next day's activities while you sleep.

TRAINING TASK

8

TASK 1

In space, astronauts take it in turns to prepare meals. Using the tables to the right, prepare five different meals for yourself and four others for a day. Try to satisfy the following needs:

Astronaut 1 – that's you and you know what you like
Astronaut 2 – does not eat meat or fish
Astronaut 3 – no special requirements
Astronaut 4 – does not eat eggs or nuts
Astronaut 5 – never feels hungry

For breakfast, each astronaut can have two foods each, and for lunch and supper two from list A and one from list B for each meal. The astronauts should also have one drink for each meal, and up to two extra if needed. During the whole day each astronaut is allowed two snacks if desired. Finally, the daily intake for each astronaut must add up to between 2,500 and 3,000 calories.

TIP
Take care to note how many packs of a food you are using – don't run out!

Foods	Packs	Calories
BREAKFAST		
Cornflakes	2	200
Rice Krispies	2	200
Bran Flakes	1	200
Bread roll and jam	4	240
Scrambled eggs	3	300
Dried fruit	3	150
SNACKS		
Peanuts	3	150
Shortbread biscuits	2	150
Dried fruit	3	150
Chocolate biscuit bar	2	150
DRINKS		
Apple	5	100
Orange	5	100
Chocolate	4	140
Strawberry	4	170
Coffee	5	20
Tea	5	20

Foods	Packs	Calories
LUNCH AND SUPPER		
LIST A		
Turkey and potatoes	2	250
Meatballs	2	350
Chicken and rice	2	350
Tuna fish and rice	1	350
Salmon and vegetables	2	350
Scrambled egg	1	300
Tomato soup and roll	2	300
Salmon and bread	2	350
Cheese and bread	3	350
Peanut butter and bread	1	350
Broccoli au gratin	3	250
Cauliflower cheese	3	250
Potatoes au gratin	5	250
Rice	5	300
LIST B		
Chocolate pudding	3	130
Jelly and strawberries	4	150
Peach rice pudding	4	150
Lemon pudding	2	370

Working Outside

■ One of the highlights of your trip will be carrying out extravehicular activity, or EVA. This means you work outside the orbiter. Your work outside will involve either capturing, repairing, and launching a satellite, launching a space probe, or testing new equipment in the weightless environment of space. In this training session you will learn how to perform these tasks.

Handy machinery

During a mission you may use the remote manipulator system, or RMS. The RMS is made up of a robotic arm, and its control system, which is in the flight deck. The RMS provides a platform on which you can stand to test new equipment or carry out repairs.

The life-support backpack contains two tanks of air – all you need for your EVA.

You can speak or signal to crew watching you as they control the RMS from the flight deck.

You use specially designed tools to carry out the repairs.

This astronaut is repairing the *Hubble Space Telescope*.

Safety straps hold your feet securely to the remote manipulator arm.

The telescope is held in place by the remote manipulator arm.

A satellite is launched from the Space Shuttle and set on its orbit in space.

Here the *Galileo* space probe is being released from *Atlantis'* payload bay.

Riding high

Satellites and space probes are launched from the payload bay into their orbits. This task is performed using the payload assistant module, PAM.

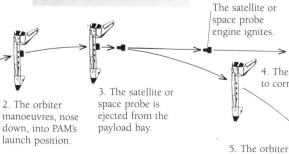

The satellite or space probe engine ignites.

The satellite or space probe is set on its orbital path.

4. The orbiter begins to correct its position.

1. A satellite or space probe is carried into space in the payload bay.

2. The orbiter manoeuvres, nose down, into PAM's launch position.

3. The satellite or space probe is ejected from the payload bay.

5. The orbiter continues along its orbital path.

PAM in action

A satellite or space probe sits in the payload bay inside a specially designed cradle. To start the launch process, the cradle opens and a table, to which the satellite is attached, starts to rotate. When the table is going fast enough, springs eject the satellite. It moves off the table at about 1 mps (3 fps).

CAPTURING A SATELLITE

This is the procedure you will follow to capture an orbiting satellite and return it to the payload bay for repair.

1. Holding the wheel-shaped capture tool, travel towards the satellite. When you are close enough, lock the tool onto the satellite.

Once the captured satellite stops spinning you can head back to the orbiter.

2. Propel the satellite towards the orbiter. Another member of the crew will be waiting on the RMS. When you are close enough, attach the special fixture on the satellite onto the RMS arm.

3. A third member of the crew in the flight deck will control the RMS to pull you and the satellite into the payload bay. Secure the satellite to a fitting inside the payload bay.

Once the satellite is secure it can either be repaired in the payload bay or taken back to Earth for refitting and relaunch on a future mission.

Pack power
The manned manoeuvring unit, or MMU, a backpack with rocket power, allows you to move around in space within about 100 m (328 ft) of the orbiter. You can steer the unit by using the hand controls at the end of each arm.

TRAINING TASK   9

Extravehicular activity
When you carry out an EVA task you will wear your EMU spacesuit. It provides all the oxygen necessary to keep you alive. The PLSS backpack contains enough oxygen for seven hours plus an extra safety supply of thirty minutes.

TASK 1
Whilst repairing a satellite, you find that it is taking longer than expected. You have already used up half your oxygen and because you are rushing you are using the remainder at a faster rate.

Each half hour's worth of remaining oxygen is now being used up in twenty minutes. You still have two and a half hours of work ahead. Can you complete it without using your safety supply?

Working in Spacelab

■ You are going to space to work. In this training session we will look at the work carried out in Spacelab. A scientific laboratory that can be fitted into the orbiter's payload bay, Spacelab is owned by the European Space Agency and used by the international community. The research you do may be on behalf of a space agency, a university research department, or a commercial group.

Active flying

This is how Spacelab looked on its first flight in *Columbia*'s payload bay in 1983. About half of the experiments on board needed an astronaut's input. This meant working in the laboratory as well as operating the controls for instruments outside on the pallets.

THE SPACE LABORATORY

Spacelab is put together on Earth and taken into space on a mission dedicated to scientific experiments. It is made up of either a long or short cabin, and a varying number of pallets, depending on the nature and number of experiments. This illustration shows the short cabin with two pallets.

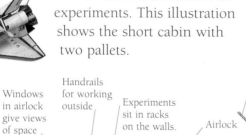

Spacelab sits in the payload bay.

Stargazing

Astronomer-astronauts can work with telescopes set on pallets in the payload bay. A telescope looking at the Sun will be used once every orbit. You will need to point the telescope at the correct part of the Sun and then instruct it to record what it sees.

Instrument for investigating Earth's magnetic field

Instrument for investigating the space environment

Telescope

Electron accelerator

Barium canisters

Solar panels

Windows in airlock give views of space

Handrails for working outside

Experiments sit in racks on the walls.

Airlock to move outside into space

Airlock

Access tunnel from orbiter to Spacelab

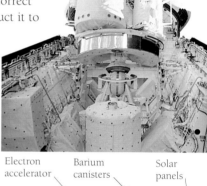

Pallets are used to support instruments exposed to the space environment.

The two pallets are linked together.

Single pallet

Pallet system

Each pallet is 3 m (10 ft) long and offers about 17 m² (56 ft²) for mounting experiments. Each has its own electrical power and cooling system.

Cabin is pressurized and air is kept fresh

Footholds in the floor

Short cabin

Insulation blanket helps keep Spacelab's temperature suitable for work in shirt sleeves

EXPERIMENTING IN SPACE

The racks on the walls of Spacelab hold many different experiments. You may find yourself making crystals to help medical research, working with a flame to find out how fire spreads in space, or even watching shrimp eggs hatch.

Astronauts work in shifts in Spacelab.

Footholds in the floor help give a sense of "up" and "down".

The lack of gravity in space causes white roots to grow above, rather than below, the soil.

These mung-bean seedlings were grown in space from seeds.

Space life
Life-science experiments look at how space affects living things. Growing plants and observing small creatures, such as frogs and jellyfish, are common areas of experimentation.

Light work
Because you are weightless you cannot weigh yourself in space. However, you can measure your mass (the amount of material of which you are made) using the odd-looking body-mass measurement device. You can compare this with your mass on Earth as recorded before the mission.

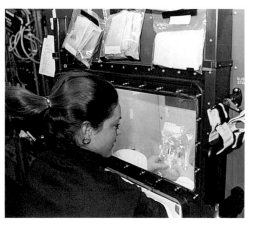

Human experiments
Scientists are continually monitoring the effects of space flight on the human body. This means you may become the subject of an experiment.

You may have to wire yourself up to a machine and record the results in Spacelab.

Glove box
If you are carrying out an experiment that is dangerous to crew members, or might interfere with other work, you can use the "glove box". This is a small, sealed unit with a see-through front that fits into the wall racks. You use it by inserting your hands into gloves in the lower wall of the unit.

TRAINING TASK
10

Growing plants
Future astronauts may grow their own food and breathe oxygen produced by plants. On Earth, a seed has everything it needs to make it grow strong and healthy. It has water to help it form roots and shoots, light to make its leaves green, and gravity to ensure its roots grow downwards to pick up lots of nutrients from the soil. In space, these things are not readily available.

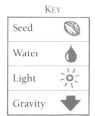

KEY
Seed	🌰
Water	💧
Light	☀
Gravity	⬇

CONDITION CHART
	Seed	Water	Light	Gravity
A	🌰	💧	☀	⬇
B	🌰	💧	☀	
C	🌰	💧		
D	🌰		☀	
E	🌰	💧		⬇

TASK 1
A seed on Earth is shown as A in the chart. Study B, C, D, and E, and work out which seed has the best chance of developing roots, shoots, and green leaves.

TASK 2
In space, water can be fed to the seed through tubes, and lamps can be used to supply the necessary light. Which condition, B, C, D, or E, do you think is possible in space?

25

Space Rendezvous

"Hello, I'm Owen Mitchell, a Space Shuttle Pilot. I've docked the orbiter with the Russian space station, and flown it home twice. In 'Module D' we'll look at space stations and your return journey."

■ The Space Shuttle orbiter is able to link up, or dock, with a space station, so in the future you may be picked for this type of mission. At present the orbiter travels to *Mir*, the Russian space station, but in a few years time this will be replaced by the bigger International Space Station. Shuttle astronauts on future trips will live and work aboard the station for weeks or months at a time.

SPACE STATIONS

Space stations are permanent orbiting spacecraft in which astronauts can carry out long-term research in a space environment. Space stations also provide humans with a space home, and the work carried out in this unique setting may one day allow astronauts to have self-sufficient lives in space.

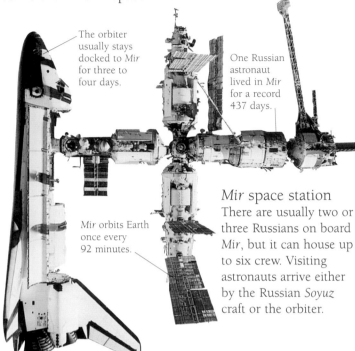

The orbiter usually stays docked to *Mir* for three to four days.

One Russian astronaut lived in *Mir* for a record 437 days.

Mir orbits Earth once every 92 minutes.

Mir space station
There are usually two or three Russians on board *Mir*, but it can house up to six crew. Visiting astronauts arrive either by the Russian *Soyuz* craft or the orbiter.

Sahara Desert
The red sand dunes in the centre of the view are one of the most prominent features of the Sahara when seen from space.

The USA at night
The bright spots in this photo are major cities in three American states: Indiana, Kentucky, and Ohio.

Great Lakes
The Great Lakes are visible from space. Lake Ontario is in the foreground.

SPACE VIEWS

One of the most fascinating things about being in space is looking out of a window at Earth. It is great fun, but what you see can also be of use on Earth. Astronauts have helped track the path of a hurricane, and spotted ash plumes from volcanoes. Astronauts who stay on *Mir* have a tradition of taking a photograph of their home from space. It is a thrill to look down on your home town from such a distance.

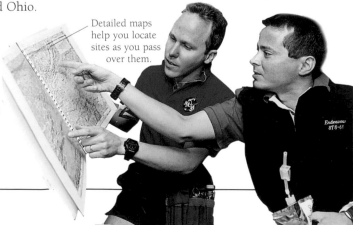

Detailed maps help you locate sites as you pass over them.

Viewing Earth

When you are in orbit in a space station, you will see some stunning views of Earth. At any one time you can see about 1,609 km (1,000 miles) of the curving Earth stretching from horizon to horizon.

TASK 1

Answer the following questions using the map to the right:
a) How many minutes will it take you to travel from A to B?
b) If Perth is on your western horizon and you look east, can you see Melbourne, yes or no?

TIP
The space station will travel over the Earth's surface at 466 kpm (290 mpm).

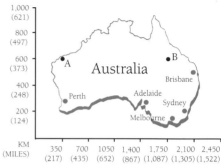

An artist's impression of the Space Shuttle docking with the International Space Station.

BUILDING A SPACE STATION

The orbiter is to play a big part in the construction and use of the International Space Station. As parts are ready, they will be transported by the orbiter and other craft, and fitted together in space. This will take 70 flights over a five-year period.

When docking two sets of hatches, one in the orbiter docking port, and one in the station, are opened.

Two astronauts welcome orbiter crew members on board *Mir*.

Stairs are not needed – astronauts simply float between levels.

Space home
The living and working modules in the International Space Station will be pressurized and have a constant supply of fresh air. Equipment will be fitted along the walls and the floors will have footholds.

International Space Station statistics:
LENGTH 88 m (290 ft)
WIDTH 108.5 m (356 ft)
WEIGHT 426 tonnes (419 tons)
CREW SIZE Up to 7

The service module provides life-support, thrusters, and habitation (toilet and hygiene) facilities.

Remote manipulator system for working outside the space station

Modules for living and working in

Laboratory

Experiments and instruments exposed to space

Pressurized experiments module

Rotating solar panels convert the Sun's energy into electricity for powering the station.

Radiators expel excessive heat

Shuttle flights
When the International Space Station is completed, the orbiter will ferry crew, fresh food, water, and supplies to it. It will also bring returning crews and waste back to Earth.

Back to Earth

■ The return journey is the trickiest part of your trip. Not only must you make a careful re-entry into Earth's atmosphere, but your landing must be perfect. Unlike an aeroplane, which can use its engines to take off again if it makes a mistake, when landing in the gliding orbiter, there is no second chance. So take your seat, strap yourself in, and get ready to learn about your return journey to Earth.

Touchdown -60 minutes
Your height is 282 km (175 miles) and your speed is 26,482 kph (16,455 mph) when you start the deorbit burn.

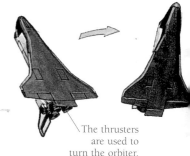

The orbital manoeuvring system engines fire to slow down the orbiter.

The thrusters are used to turn the orbiter.

PREPARING TO LAND

About an hour and a half before touchdown you should drink a saline solution. This is a precaution against fainting, which can be brought on by dehydration during your descent. You then put on your partial pressure suit and strap yourself into your seat.

Pilot power
The commander (left) and the pilot (right) sit at the flight deck for the return journey. Apart from a brief communications blackout they are in contact with ground staff for all of the return flight.

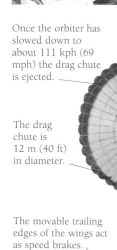

Protective shell
There are about 32,000 tiles covering the outer surface of the orbiter. They stop the orbiter from burning up as it re-enters Earth's atmosphere.

Once the orbiter has slowed down to about 111 kph (69 mph) the drag chute is ejected.

The rudder splits, keeping the orbiter stable as it slows down.

The drag chute is 12 m (40 ft) in diameter.

The movable trailing edges of the wings act as speed brakes.

Silicon city
There are three types of tile: white, grey, and black, each offering a different level of protection. Over 20,000 of the tiles are black and made from silica like this one. The coating protects the orbiter from temperatures of up to 1,260 °C (2,300 °F).

G9428 001
-TPS-3 B55857
V070-191011-072

Stopping the orbiter
The orbiter has special carbon breaks, a rudder that splits in two, and movable trailing edges, called elevons, on the wings to help it slow down. You will use a drag chute once you are on the runway to reduce the speed of the orbiter.

Touchdown -14 seconds
Just 335 m (1,099 ft) from the runway, you lower the landing gear and pull the orbiter's nose up to a 1.5° slope.

Touchdown
You touch down on the 4,572-m- (15,000-ft-) long runway doing 346 kph (215 mph).

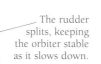

Touchdown -30 minutes
Your height is 122 km (76 miles) and your speed is 25,877.5 kph (16,083 mph) as you re-enter Earth's atmosphere.

Communication breakdown

When you re-enter Earth's atmosphere you will experience a communications blackout as the radio links with Earth are severed. This happens between approximately 85 km (53 miles) and 53 km (33 miles) above Earth. As you return to Earth, three main things are changing: your speed, height, and direction.

HEIGHT

Deorbit burn

100 km (62 miles)

Orbiter enters Earth's atmosphere

80 km (50 miles)

Communications blackout begins

60 km (37 miles)

40 km (25 miles)

Communications blackout ends

Touchdown

FLIGHT PROFILE

TASK 1
If the orbiter is losing height at a rate of 2 km (1.25 miles) per minute, how long does the blackout last?

Burning up
The angle of re-entry into Earth's atmosphere must be carefully calculated. If it is too steep, friction caused by air makes the orbiter burn up. If it is too shallow, the orbiter skips out of the atmosphere.

Touchdown -20 minutes
When your height is 70 km (43.5 miles) the outside of the orbiter heats up to 1,371 °C (2,500 °F).

Touchdown -86 seconds
The final landing stage begins when you are 12 km (7.5 miles) from the runway and your speed is 682 kph (424 mph).

Space plane to glider
As soon as the orbiter enters Earth's atmosphere air currents flow over the craft. The delta wings that were redundant in space now help the craft to glide as it descends.

Touchdown -32 seconds
When you are just 3.2 km (2 miles) from the runway and your speed is 565 kph (351 mph) you start to descend at a 20° slope.

Coming in to land
During the approach and landing stage you drop at the rate of 3 km (1.9 miles) per minute. This is 20 times greater than the speed of an airliner's descent.

Touchdown -17 seconds
You are 1,079 m (3,540 ft) from the runway and travelling at a speed of 496 kph (308 mph).

MAXIMUM CAPACITY 4 PERSONS OR 1000 LBS.

Feet on the ground
Although you can unfasten your straps once the orbiter is on the ground, you have to wait for an hour before you can get out. The orbiter must cool down and be cleared of explosive or toxic (poisonous) fumes before you leave the craft.

NASA

United States

The body flap between the wings and below the main engines is used to position the orbiter during flight in Earth's atmosphere.

Index

Picture credits
Every effort has been made to trace the copyright holders. Dorling Kindersley apologizes for any unintentional omissions and would be pleased, in such cases, to add an acknowledgment in future editions. The publisher would like to thank the following individuals, companies, and picture libraries for their kind permission to reproduce their photographs:
Genesis Space Photo Library/NASA: 2, 9cr, 12t, 12br, 13b, 16cr, 18cl, 21tl, 21tc, 21cl, 21c, 26b, 27tl; © NASA: 4b, 9c, 9br, 12bl, 13tl, 13tr, 13c, 14t, 14cl, 14c, 14bl, 14br, 15t, 15tl-tr, 15c, 15bl, 16cl, 16b, 17t, 17b, 19t, 19c, 20t, 21tcl, 21cr, 22cl, 22c, 22cr, 23tbl, 23c, 23b, 24t, 24cr, 25tl, 25tr, 25c, 25cbl, 25cbr, 26t, 26cl, 26c, 26cr, 27c, 27b, 28cl, 28cr, 28b, 29c, 29crb, 29b; Popperfoto/Reuter: 12c; Science Photo Library/NASA: 15br, 16t, 23tr, 23cl, 27cr; Science & Society Picture Library/NASA: 30. The Telegraph Colour Library: 17c, 23tl. Jacket: NASA: front bl, cr; Science Museum: front cl.

⚡ TRAINING TASK ANSWERS

Training Task 1
Task 1 – 1 hr 30 mins or 90 mins
Task 2 – 16 times

Training Task 2
Task 1 – **a**
Task 2 – slowest
Task 3 – eccentric

Training Task 3
Task 1
a Yaw, **b** Roll, **c** Pitch, **d** Pitch
e Pitch, **f** Roll, **g** Roll and pitch
h Pitch

Training Task 4
Task 1
1) **b**, 2) **a**

Training Task 5
Task 1
b – 44 kg (98 lb)
c – 25 kg (55 lb)
d – 16 kg (35 lb)

Training Task 6
Task 1
1) **b**
2) The blood would form a sphere and start to float around the orbiter.
3) **c**

Training Task 7
Task 1
a) 10,000 times louder
b) 100,000 times louder
Task 2 – G

Training Task 8
Task 1
Check your own answer. All menus, including that for astronaut 5, must add up to between 2,500 and 3,000 calories. Add up the number of times you have used each food and make sure it does not exceed the number of packs listed.

Training Task 9
Task 1 – No

Training Task 10
Task 1 – B
Task 2 – B. Everything except gravity can be provided in space.

Training Task 11
Task 1
a) 3 minutes
b) Yes

Training Task 12
Task 1 – 16 minutes

The module instructors featured in this book are fictitious and any resemblance to actual persons, living or dead, is purely coincidental.

Acknowledgments
Dorling Kindersley would like to thank the following people for contributing to the production of this book: Stephen Bull for the training task illustrations, David Hughes and Jane Earland for consultancy, and Lia Foa.